HOW TO USE HAWAIIAN FRUIT

by Agnes B. Alexander

Cover illustration by
Jan Moon

Text illustrations by
William D. Brooks

Published by the
Petroglyph Press

HOW TO USE HAWAIIAN FRUIT
Copyright 1974, 1999
Published by the Petroglyph Press, Ltd.
160 Kamehameha Avenue
Hilo, Hawai'i 96720
Voice 888-666-8644 / Fax 808-935-1553
BBInfo@BasicallyBooks.com
www.BasicallyBooks.com

ISBN 0-912180-53-6

Cover design and cover printing by Don O'Reilly,
Hilo Bay Printing.

First Edition ~ 1974
Second Edition
1st Printing ~ January 1999
2nd Printing ~ October 2001

How to Use Hawaiian Fruit was originally published in 1912 in Honolulu, where the author once owned a popular restaurant. Petroglyph Press added illustrations by William D. Brooks in 1974 and in 1998 reset the type and added a new cover design by Jan Moon, but has kept the flavor of the original cookbook. Almost all of the recipes are very simple and call for basic ingredients, some no longer common in this day of packaged foods. But note, with amusement, the recipe for Lomilomi Salmon on page 73 which calls for "5 cents onions and 5 cents tomatoes". How times have changed!

INTRODUCTION

The measures used in this book are c, level cup, or ½ pint; tsp, level teaspoon, and tbsp, or level tablespoon.

With the exception of the *ohelo*, the fruit and food products selected are all found in our market.

It has been found that fruits retain their flavor better, when cooked at a low temperature.

In canning fruit, sterilize bottles, covers, and rubbers and fill bottles quickly. In this way all bacteria germs are killed, and the fruit should keep indefinitely.

When fruit ripens, the starch composition is changed into levulose, or glucose substance, which is more easily assimilated. The tannin and vegetable acids are also changed, and the fruit becomes less stringent. If the fruit is over ripe, bacteria have been introduced.

The original fruit and food products of Hawaii nei, which were found by the first discoverers, are nine. These are the taro, or *kalo*; the banana, or *maia*; the breadfruit, or *ulu*; the mountain apple, or *ohia*; the coconut, or *niu*; the *akala*, the *ohelo*, the wild strawberry, or *ohelo papa*; and the sugar cane, or *ko*.

Some of these may have been brought to our shores by the aboriginal settlers, in their canoes; while the currents might have brought the coconut, and birds carried the seeds of the berries.

MIXED FRUIT

CONTENTS

AVOCADO, OR "ALLIGATOR PEAR"

This fruit is rich in fat, containing as high as 10 per cent. There are many varieties which vary when ripe in color from green to purple. It is usually served as a salad.

AVOCADO IN SOUP AND STEW.

Scoop out the avocado pulp with a spoon and drop in soup just before serving. Serve in the same way in stew.

AVOCADO COCKTAIL.

1. Cut fruit into dice, or small pieces, and serve in glasses with cocktail sauce made of tomato catsup and lemon juice, and season with salt and pepper. Serve with chopped ice.

2. Mash the fruit and beat it up with the cocktail sauce. Serve in the same manner.

AVOCADO SALAD.

In preparing avocado salad, one may either cut it in dice, scoop it with a spoon, mash it and mix with the dressing, or serve it in the half skin and cover with the dressing.

AVOCADO VEGETABLE SALAD.

Combine sliced avocado, cucumber, tomato, and a few shreds of Chili pepper and serve with mayonnaise or French dressing.

AVOCADO SALAD WITH STRING BEANS.

Combine equal parts of cooked string beans and avocado and serve with French dressing.

AVOCADO SALAD IN TOMATOES.

Press the avocado through a potato press, mix with mayonnaise dressing, strongly seasoned with mustard, and fill peeled and partly scooped out tomatoes with the avocado.

AVOCADO SERVED IN SKIN.

Cut avocado in half, remove seed, and serve with one of these five dressings; l, sugar; 2, salt and vinegar; 3, tomato catsup; 4, Durkee's salad dressing; 5, French dressing.

Or partly fill the cavity with sliced celery and sliced walnuts, and cover with mayonnaise.

AVOCADO ASPIC SALAD-1.

½ box gelatine   2 c mashed avocado pulp
½ cup cold water   Juice ½ lemon
1 cup boiling water   Salt, cayenne

Soak the gelatine in the cold water ½ hour. Dissolve in the boiling water. Strain and add the fruit pulp which has been flavored with salt, cayenne, and the lemon juice. Place on ice to harden. Serve with mayonnaise.

AVOCADO ASPIC SALAD-2.

½ box gelatine   Juice of 2 lemons
½ c cold water   2 tbsp sugar
1½ c boiling water   2 tomatoes
1 good sized avocado

Soak the gelatine in the cold water ½ hr. Dissolve in the boiling water. Strain and add sugar and lemon juice. Cut the avocado and tomato in cubes. Place a mold in ice water and pour in a layer of jelly, then avocado and tomato. Add a little jelly to keep the fruit in place, and when set, add more and more fruit. Place on ice to harden. Serve with mayonnaise. This and the preceding salad may be molded in individual molds and served in nests of lettuce leaves.

AVOCADO AND PINEAPPLE SALAD.

Slice the avocado in small pieces and mix with cubed pineapple. Serve with French or mayonnaise dressing.

AVOCADO SALAD DRESSING.

Press avocado through a sieve or potato press. Mix vinegar, red pepper, salt, mustard, a little lemon, or lime juice, as for salad dressing. Add equal parts of oil and avocado pulp, adding the oil gradually to the other ingredients.

AVOCADO SANDWICHES-1

Mash the pulp of the fruit and season with salt and pepper. Spread this between thin slices of bread.

AVOCADO SANDWICHES-2.

Slice the fruit with chopped chili peppers or cayenne, and place between thin slices of buttered bread.

AVOCADO ICE CREAM.

Yolks 5 eggs   2 c sugar
1 qt. milk   4 medium sized avocados
Green maraschino cherries
Almond and vanilla extracts

Make a boiled custard with the milk, eggs, and 1 c sugar, and flavor with vanilla. Mash the fruit to a pulp with 1 c sugar and flavor with almond extract. When the custard is cool, add the fruit and freeze. Serve in mounds with a green maraschino cherry on top of each mound.

BANANA

The banana contains a greater percent of food material than any other fruit, being chiefly starch and sugar. There are two general types, the Chinese and like varieties, only suitable for eating raw and frying, while the plantain or cooking banana may be boiled, baked, or fried. The strings which cling to the fruit are the only indigestible part, and should be removed before eating. The unripe fruit may be dried and ground into meal. Some use it as a beverage, like coffee.

BANANAS SIMPLE.

Slice and serve with lemon or China orange juice, or with sugar and cream. They may also be sliced into porridge.

DRIED BANANAS.

Peel bananas and cut in thin flakes, or press through a sieve. Dry in the sun, or hot attic on screened dishes. When dry, keep in a dry, dark cool place.

BAKED BANANAS-1,

Bake bananas in their skins, or peel not too ripe ones. Add a little salt and put in same pan with a roast ½ hour before serving. Turn once or twice. Serve as a vegetable.

BAKED BANANAS-2.

Peel bananas and place in a baking pan. Sprinkle with sugar, lemon juice, and a little cinnamon. Bake ½ hour and serve as a vegetable.

BAKED BANANAS—3.

Bake bananas in their skins 45 minutes in a hot oven. Peel and place on a lettuce leaf with whipped cream on top. Serve cold as an entrée.

Or bake and serve in the same manner with grated coconut on top of the bananas.

BANANAS AND DATES.

Serve finely sliced dates and bananas with whipped or plain cream, or plain custard.

FRIED BANANAS.

Slice bananas in 2 or 3 strips and fry in half butter and half lard.

BANANA CROQUETTES.

Cut bananas as to fry, and dip in egg and bread crumbs. Fry in deep fat and drain on paper.

DISGUISED BANANAS.

Peel and cut baking bananas each into 5 or 6 pieces. Add sugar, bits of butter, cinnamon, a little fruit juice, or tamarinds and water, or just water, and bake slowly in a covered pan for 3 or 4 hours, or quickly in an uncovered pan.

STEWED BANANAS-1.

Cut bananas in small pieces, add sugar and a few tamarinds; or China orange, lime, or lemon juice and sugar, and simmer slowly in a very little water. Serve as a fruit course.

STEWED BANANAS-2.

Peel and cut bananas in halves. Cover with a little water and stew ½ hr. Serve as a vegetable.

BANANA FRITTERS.

3 bananas	1 c flour
¼ c milk	1 tbsp powdered sugar
¼ tsp salt	2 tsp baking powder
1 egg	

Mix and sift dry ingredients. Beat egg, add milk, and combine mixtures. Add sliced or mashed bananas. Fry in deep fat and drain on paper.

BANANA SOUP.

1 pt. milk	¾ c mashed and strained banana
1 tbsp butter	1 tbsp flour
¼ tsp salt	Pepper

Melt the butter and add flour, mixing thoroughly. Add milk slowly, stirring constantly. When it boils, add the bananas and cook 5 minutes. Serve hot with crackers. For this use the Chinese or cold baked banana.

BAKED BANANA DUMPLINGS.

1 c flour	2 tsp baking powder
1 heaping tbsp butter	½ tsp salt
½ c water	Banana and seasoning

Rub together the butter and flour sifted with the baking powder and salt. Add about ½ c water and mix lightly with spoon. Divide into 4 balls and roll out lightly and quickly to the size of large saucers. Put into the middle of each round minced banana, sugar, butter, and lemon extract, and bring paste to top, wetting the edges and leaving opening for steam to escape. Serve hot with sauce made of 4 tbsp sugar, 2 tbsp butter, cinnamon and a pinch of salt creamed together. Add ½ c boiling water.

BANANA SALAD-1.

Make nests of lettuce leaves on individual plates. Slice bananas and lay the slices one on the other in the form of a circle within the lettuce, leaving the center empty. In this place a spoonful of mayonnaise and on top lay a candied cherry or a strawberry.

BANANA SALAD-2.

Slice bananas in halves lengthwise. Lay them on lettuce leaves, and place on top of each half, chopped walnut; then a narrow layer of mayonnaise.

BANANA SALAD-3.

Slice bananas, celery and walnuts. Mix with salt and mayonnaise, and serve in banana boats made of the half skins with lettuce leaves for sails. Place in nest of lettuce with 2 or 3 stuffed olives.

BANANA SALAD-4.

Serve cold baked bananas on lettuce leaves with mayonnaise or boiled dressing.

BANANA PIE-1.

Slice bananas thin, add sugar, nutmeg, butter and guava jelly, or ½ c finely chopped pineapple pickle, or a larger quantity of chopped fresh pineapple. Bake slowly between two crusts.

BANANA PIE-2.

Slice thin, enough bananas for one pie. Mix ¼ c sugar, little salt, 1 tbsp China orange or lemon juice, and spread half of this on the pie crust. Dot with tsp butter, then put in a layer of bananas and repeat. Cover with a top crust and bake quickly.

BANANA HARD SAUCE.

Cream together ½ c sugar and ¼ c butter, beating well; then add ½ c banana pulp and beat. The white of one egg may be added before the fruit.

BANANA PLAIN SAUCE.

1 tbsp butter	1 tbsp flour
6 tbsp sugar	½ c orange juice
½ c banana pulp	

Melt the butter, add flour, mixing thoroughly, then the sugar and fruit. Stir and serve while still foaming. This may be used with any plain pudding.

BANANA BROWN BETTY.

Put a layer of sliced bananas in buttered dish. Add sugar, butter, and a little chopped pineapple pickle with syrup, or orange marmalade, or guava jelly; then layer of bread crumbs sprinkled with cinnamon; then layer of bananas, etc., making the last layer of crumbs. Bake slowly and serve hot with cream. If fresh pineapple and juice are used, the cream may be dispensed with. The whipped whites of 2 eggs with 2 tbsp sugar, put lightly on top, and browned quickly in the oven after the pudding is cooked, is an improvement.

BANANA BLANC MANGE.

1 qt. milk	½ c sugar
3 tbsp cornstarch	Salt
½ c cold water	3 large bananas
White 1 egg	

Dissolve cornstarch in cold water, add to boiling milk, and when it thickens add sugar and bananas, which have been pressed through a sieve. Then add the stiffly whipped white of egg.

ROLY-POLY BANANA PUDDING.

1 c flour	½ c cold suet
⅔ c ice water	½ tsp salt
2 tsp baking powder	

Sift flour, baking powder, and salt. Chop into this scant ½ c suet and mix quickly with ice water. Roll out thin and long and spread with thin slices of banana and sprinkle with lemon juice. Roll up like a jelly roll and bake in hot oven ¾ hr. Serve hot with sauce.

BANANA PUDDING.

1 qt. milk	2 or 3 bananas
Yolks 2 eggs	2 tbsp cornstarch
2 tbsp powdered sugar	Salt

Boil the milk and stir in the beaten yolks of eggs, sugar, cornstarch dissolved in a little cold milk, and pinch of salt. Remove from the fire when it thickens. Take baked bananas and slice in a pudding dish. Add powdered sugar and cover with the custard. Put in the oven and brown.

BANANA GELATINE.

½ box gelatine	½ c cold water
½ c sugar	1 c boiling water
3 Chinese bananas pressed through sieve	

Soak the gelatine in the cold water. Boil the sugar and 1 c water, add the gelatine and orange juice. Place in a mold and set on ice to harden. Serve with cream.

BANANA SPONGE.

1 tbsp gelatine	¼ c cold water
1 c banana pulp	Juice ½ lemon
Whites 2 eggs	½ c sugar
1 doz. chopped nuts	

Soak gelatine in cold water 5 minutes. Stir and cook banana pulp, lemon juice, and sugar until boiling, and add gelatine. Place in ice water, and when it begins to set, fold in the stiffly beaten egg whites. Pour into mold and place on ice. Sprinkle with the chopped nuts.

BANANA MARMALADE.

4 bananas   1 c sugar
1 c water   Juice 1 lemon

Press bananas through potato press, and add water. Place on the fire, and when it boils, add sugar and lemon juice, stirring. Cook 20 minutes. Serve with hot cakes or waffles.

BANANA WITH JUNKET.

Fill sherbet glasses two-thirds full of plain milk junket made from junket tablets. Crush and sweeten bananas and place on top of junket. Cream may be mixed with the banana.

BANANA CUSTARD PIE.

2 bananas   2 eggs
½ c sugar   1 pt. hot milk
Salt   Cinnamon or vanilla

Mash and strain the bananas, add well beaten egg, and whip together until light and frothy. Add sugar, hot milk, salt, and cinnamon or vanilla. Bake in one crust.

BANANA FILLING FOR CAKE.

6 bananas   Whites 2 eggs
¼ c sugar   Juice 1 orange and grated rind

Mix the sugar and grated rind. Add the bananas mashed. Whip the egg whites and beat mixture into it. Whipped cream may be used in place of eggs.

BANANA WHIP.

To 3 Chinese bananas allow the white of 1 egg. Press the banana through a potato press. Sweeten with powdered sugar. Add this to the well whipped egg white. Serve in sherbet cups, ice cold. Whipped cream may be used in place of the egg white.

BANANA SOUFFLÉ.

1 c mashed bananas   Whites 4 eggs
Powdered sugar   Salt

Whip the egg whites with a pinch of salt, and beat into them the sweetened banana pulp. Turn into a buttered pudding dish, or individual molds in pan of hot water, and bake in slow oven until firm.

BANANA CREAM.

¼ box gelatine	Juice 1 lemon
¼ c cold water	4 bananas mashed
1 c cream	¼ c sugar

Mix the bananas with the sugar. Whip the cream, and stir in the bananas and lemon juice. Soak the gelatine in the cold water 5 minutes, and dissolve by setting dish in hot water. Strain and mix with the banana. Place in a mold to harden on ice.

FROZEN BANANA CUSTARD.

4 tbsp sugar	4 bananas
3 eggs	3 c milk
3 tbsp cold water	1 c cream
1 tsp gelatine	

Soak gelatine in the cold water. Place milk on stove, and add sugar and eggs, stirring. Add gelatine, and stir until it thickens. Cool, and add bananas pressed through sieve, and fold in the whipped cream. Freeze.

BANANA SHERBET.

1 pt. water	1 c banana pulp
1 c sugar	Juice 1 lemon
1 tsp gelatine	

Boil the water and sugar 20 minutes. Soak the gelatine in little cold water 5 minutes. Add it and the lemon juice to the syrup. Strain. When cool add the banana and freeze.

BANANA ICE CREAM.

To 1 c mashed banana pulp add the juice of ½ lemon. Mix this with 1 qt. of any plain ice cream when partly frozen.

BREADFRUIT

This fruit, as it is called, is used only as a vegetable, and when properly baked, it is somewhat similar to the sweet potato. It is stringy only when undercooked.

TO RIPEN BREADFRUIT.

Remove the core from slightly green breadfruit, and put in its place 1 tbsp rock salt. In about 2 days it will be ripe.

BAKED BREADFRUIT-1.

Take the fruit when it is just turning soft, but not too soft nor too hard. The whole secret is in getting it at the right stage. Bake in a hot oven for 1 hr. Cut in half, remove the seed and break in individual pieces. Serve with butter.

BAKED BREADFRUIT-2.

Cut the stem and core of a breadfruit out together. Put butter and salt in the cavity, replace the stem and bake for 1 hour.

BOILED BREADFRUIT.

Cut the skin from a ripe breadfruit. Tie pulp in a cotton bag, and place bag in boiling water. Boil for 40 min. to 1 hr., according to size of breadfruit. When done, roll out in a dish, slice, and serve with melted butter.

BREADFRUIT WITH STEW.

Baked, or boiled breadfruit may be served in stew in place of potatoes.

BREADFRUIT PUDDING-"PAPAIEE."

Take ripe breadfruit, scrape out the pulp and mix well with the milk, and pressed juice from a grated coconut. Add ½ tsp salt and 1 tbsp sugar. Bake ½ hr. to 1 hr.

CHINESE ORANGE

The China orange is the small acid orange. It may be used in place of lemon, and is very good cut in half and served with papaya for breakfast.

CHINA ORANGE DRINK.
Make as lemonade, and sweeten to taste.

CHINA ORANGE SERVED WITH TEA.
With a sharp knife cut the oranges crosswise into thin slices. Serve these with tea in place of lemon.

CHINA ORANGE PUNCH.
3 c strong tea   ½ tsp almond extract
1 c China orange juice   Mineral or plain water
Sugar

Dilute tea and orange juice with water and sweeten to taste. The water may be part mineral. Eight stewed peach leaves may be used in place of the almond extract.

CHINA ORANGE MARMALADE.
Clean fruit carefully. Squeeze seeds from oranges and save juice. Strain juice free from seeds. Put skins and pulp through a meat chopper, and place in kettle. Add juice and boil. Then add measure for measure of sugar, stirring until sugar is dissolved, and boil until it becomes glassy.

Or pour boiling water over oranges and let stand over night. Then peel skins off and cut into strips. Free pulp from seeds, add skins, and boil slowly. Add equal measure sugar and boil until thick and glassy.

CHINA ORANGE AND PAPAYA MARMALADE-1.
To 1 measure papaya allow ½ measure China oranges. Wash oranges well. Squeeze out seeds and juice. Put skins through a meat chopper and add to the strained juice. Add papaya pulp cut in small pieces and boil all together; then add as much sugar as pulp. Boil again for 15 or 20 minutes.

CHINA ORANGE AND PAPAYA MARMALADE-2.
To 6 c papaya cut in small pieces add ⅓ c China orange juice. Boil 15 minutes and add ½ as much sugar as pulp. Boil again for 15 or 20 minutes.

CHINA ORANGE AND PAPAYA MARMALADE-3.
3 qts. China oranges   5 small papayas
Soak the skins of ½ the oranges over night, or boil, changing the water three times. Squeeze the juice from the oranges. To 1 c of juice add 2 c sugar. Cook the orange juice and skins with sugar until stringy, then add papaya which has been cooked until soft.

CHINA ORANGE PRESERVES.
Pour over oranges a boiling salt brine and leave over night. In morning drain and pour on fresh water. Wipe oranges and open at stem end with small silver knife. Remove seeds with out breaking the fruit. Make a syrup of 4 parts sugar to 1 part water. Boil and drop oranges in. Boil hard for ¾ to 1 hr.

CHINA ORANGE PIE.
Juice and pulp 7 oranges 1 c sugar
1 egg   1 tbsp cornstarch
½ c boiling water
Peel oranges and remove seeds, saving juice. Dissolve cornstarch in a little cold water and cook in the boiling water, adding sugar, oranges, and beaten egg. Pour into a pie crust and bake. Strips of pie crust may be placed on the top, or a meringue added when the pie is done, and browned in the oven.

CHINA ORANGE GELATINE.
½ c cold water   1½ c boiling water
½ c orange juice   ¾ c sugar
½ box gelatine   ½ c China orange juice
Soak the gelatine in cold water 5 minutes, and dissolve in the boiling water. Add sugar and stir until dissolved and cool. Then add fruit juice and strain through cloth into mold. Set on ice to harden.

CHINA ORANGE SPONGE.

1-3 box gelatine	1 c sugar
1-3 c cold water	Whites 3 eggs
1 c boiling water	Juice 3 China oranges

Soak gelatine in cold water 5 minutes and dissolve in boiling water. Add sugar and strain. Place in ice water, and when it thickens beat in the stiffly whipped egg whites, and beat with a Dover egg beater until light and spongy. Put lightly into glass dish, or place in mold on ice.

CHINA ORANGE CUSTARD.

3 c water	2 eggs
3 large China oranges	1 c sugar
2 tbsp cornstarch	

Boil water and add moistened cornstarch. Stir until it thickens; then add yolks of eggs beaten with the sugar and orange juice. Stir one minute, and pour into glass dish, or sherbet glasses. Make a meringue of the egg whites, and flavor with juice, and grated skin of a China orange.

CHINA ORANGE SHERBET.

White 1 egg	1 qt. water
2 c sugar	½ c China orange juice

Boil the sugar and water 20 minutes, stirring until the sugar is dissolved. Cool and add China orange juice. Freeze, and when partly frozen add the beaten white of egg.

COCONUT

The coconut may be used both in the ripe and unripe state. It ranks high in food value, containing 25% fat and 14% starch. The milk is a refreshing drink, and may be used in place of cow's milk. To extract the milk from old coconuts, grate coconut and pour over it boiling water, 1 pt. to a coconut. Stir well, strain and squeeze through a cotton cloth.

To keep grated coconut, sprinkle it with sugar and place on a sieve in the sun to dry.

"KOELE PALAU."

Mash baked or boiled sweet potato and pour over an equal measure of grated coconut milk. Sweeten to taste. Serve hot.

"HAUPIA."

Take equal measures of pia or cornstarch and grated coconut. Soften the pia with water. Heat the sweetened coconut milk and pour in pia gradually. Stir only one way, until very smooth. Serve cold. Pia is an Hawaiian starch.

"KULOZO."

Mix well 1½ lbs. grated raw taro with the milk and strained juice of a grated coconut. Add a little salt and enough sugar to take away the sharp biting taste of the taro, about 2 tbsp. Bake slowly for 2 or 3 hours. Sweet potatoes can be used instead of taro.

COCONUT AND CORN PUDDING.

1 doz. ears corn	1 grated coconut
1 pt. milk	4 eggs
1 tsp salt	¼ tsp black pepper

Cut corn from the cob, and mix with coconut and beaten yolk of eggs. Add salt, pepper, and milk. Stir in the beaten whites of eggs and bake in a pudding dish 1 hr. Serve as a vegetable.

COCONUT PUDDING-1.

2 eggs   3 tbsp sugar
2 tbsp cornstarch   1½ c cow's or coconut milk
½ c grated coconut

Put milk on stove, add sugar, stirring. When it begins to boil, add cornstarch which has been dissolved in a little cold milk. Stir, and as it thickens add the beaten white of eggs. Remove, and when partly cool, add grated coconut. Mix well and turn into a mold. Make a custard with the yolks of eggs, 1 c milk and 2 tbsp sugar. Pour custard around pudding when served.

COCONUT PUDDING-2.

1 pt. milk   2 tbsp cornstarch
2 tbsp sugar   Yolks 2 eggs
Pinch salt

Heat part of the milk, salt, and sugar. Dissolve the cornstarch in the remainder of the milk. Add to it the heated milk, and when cooked add the beaten yolks of eggs, and ¼ c grated coconut. Flavor with vanilla. Beat whites stiff, sweeten and flavor with lemon; spread on top and brown quickly in hot oven.

YOUNG COCONUT.

Serve young coconut in the shell, or scoop out and serve in a dish. Stewed mulberries may be added to the coconut.

COCONUT PIE.

1 pt. coconut or cow's milk   2 eggs
1 c grated coconut   3 tbsp sugar
¼ tsp salt

Bake pie crust. Beat yolks of eggs ; add sugar, milk, coconut, and salt. Make 1 thick or 2 thin pies. Bake 30 minutes. Add meringue of white of eggs beaten with 2 tbsp sugar and flavored with vanilla. Brown slightly in oven.

COCONUT COOKIES.

1 c grated coconut   ¾ c butter
1 ½ c sugar   2 c flour
2 eggs   2 tsp baking powder
½ c milk   Grated rind 1 lemon or ½ tsp extract

Mix butter, sugar, beaten eggs, and milk. Sift flour, baking powder, and salt. Add this to the first mixture and grate the lemon over it. Stir in the coconut, and add enough flour to roll out; or drop in rough cakes on greased pans. Bake 15 minutes in moderate oven.

COCONUT FILLING OR FROSTING.

½ c grated coconut   ¾ c powdered sugar
Whites of 2 eggs
Beat the eggs stiff, add sugar; beat, and add coconut.

COCONUT SAUCE.

1 c grated coconut   ½ c butter
1 egg   1 c sugar
Cream butter and sugar, add beaten egg, and beat well, adding coconut.

COCONUT DROPS.

1½ c sugar   2 c grated coconut
White 1 egg   Salt
Boil the sugar and coconut until it becomes sticky. When cool, add the stiffly beaten white of an egg to which a little salt has been added. Beat together and drop from spoon on buttered pans, and bake in moderate oven 15 minutes or until brown. Take pan from oven, and let cool before taking up the drops.

COCONUT FONDANT CANDY.

Dissolve 1 c sugar in 1 c water. Add tsp cream of tartar. Boil to soft ball stage, taking care that the syrup does not form crystals on the side of the pan. Grate coconut and mix with the fondant as much as it will hold well. Spread on buttered plates, and cut into 1 inch squares. Leave to harden.

COCONUT CREAM CANDY.

Boil ½ c milk with 3 c brown sugar, add 2 c grated coconut, and cook until it strings. Flavor with vanilla. Set in pan of cold water to cool, and beat hard until creamy. Drop on paraffin or buttered paper.

COCONUT MOUSSE.

1 c grated coconut	3 tbs orange juice
1 tsp gelatine	1 pt. cream
½ c boiling milk	2 tbsp cold water

Soak gelatin in cold water and dissolve in boiling milk. When cool, add coconut, orange juice, and whipped cream. Place in mold or freezer, and pack in salt and ice for 4 hours.

Coconuts

COFFEE

The Hawaiian Kona coffee is considered by some the best in the world. It should be kept in an airtight container and freshly ground before using.

COFFEE BEVERAGE.

Kona coffee has been found best made in this way. Allow 1 tbsp coffee to 1 c water. Measure and leave over night. In morning simmer, but not boil, 20 minutes. No egg is needed to clear.

COFFEE JELLY.

½ box gelatine   3 c black coffee
½ c cold water   ¾ c sugar

Soak the gelatine in the cold water and dissolve in the hot coffee. Add sugar and stir until it dissolves. Strain and turn into a mold. Serve with whipped cream.

COFFEE JUNKET.

2 tbsp sugar   1 c strong coffee
3 c milk   1 tbsp cold water
Junket tablet

Crush and dissolve junket tablet in tbsp water. Heat milk, add sugar and coffee. Take from fire and when lukewarm add junket tablet. Serve with cream.

COFFEE SPONGE.

1 c black coffee   1 c sugar
½ box gelatine   ¾ c cold water
Whites of 3 eggs

Soak gelatine in cold water 5 minutes. Dissolve in boiling coffee and add sugar, stirring. Strain and set in ice water. When it begins to thicken, beat the whites of eggs to a stiff froth, and whip the jelly into them, a spoonful at a time. Beat with an egg beater. Pour into mold and place on ice to harden. Serve with cream.

COFFEE SPANISH CREAM.

¼ box gelatine	2 eggs
1½ c strong coffee	Pinch salt
⅔ c sugar	1 tsp vanilla
⅔ c milk	

Soak gelatine in ½ c cold coffee 5 minutes. Boil 1 c coffee and milk. Add beaten egg yolks and cook 3 minutes; then add sugar, dissolved gelatine and salt. When cold, beat in stiffly whipped whites of eggs, flavor and place in mold on ice. Serve with cream.

COFFEE CHARLOTTE RUSSE.

¾ box gelatine	½ c clear coffee
1 c milk	½ c sugar
¼ tsp salt	2 tbsp powdered sugar
3 c cream	Lady fingers

Soak gelatine in cold milk and dissolve over hot water, adding ½ c sugar. Beat egg yolks, add powdered sugar, salt, scalded milk, and coffee. Strain and set in ice water. Stir until it thickens; then fold in the whipped cream. Turn into mold lined with lady fingers.

COFFEE MOUSSE.

1 tbsp gelatine	⅔ c sugar
1 pt. whipped cream	3 c coffee
Whites of 3 eggs	

Soak gelatine in a little cold water 5 minutes and dissolve in the hot coffee. Cool. Beat all together and place in freezer or mold packed in equal parts of salt and ice for 4 hrs.

COFFEE BAVARIAN CREAM.

⅔ c powdered sugar	1 pt. cream
1½ c boiling coffee	½ box gelatine
½ c cold water	

Soak gelatine in the cold water, and add to it the hot coffee. Sweeten, turn into a mold in ice water, and when it thickens fold in whipped cream and place on ice to harden.

COFFEE ICE CREAM.

1 qt. cream   1 c sugar

½ c black coffee

Bought bottled cream can be diluted with milk. Scald the cream and sugar in a double boiler, and when cold add the coffee. Freeze.

FIG
The fig is considered one of the most healthful fruits known.

FIGS SIMPLEST.
Peel figs and slice. Eat with sugar and cream.

Or cut 2 figs in halves, and place the four halves on a plate in form of four leafed clover. Sugar, and put a maraschino cherry in each. Strawberries or seeded grapes could be used in place of cherries. Eat with spoon.

FIG SALAD.
Peel and slice figs and mix with shredded pineapple. Serve on lettuce leaves with mayonnaise.

Or peel and fill figs with chopped nuts and cover with mayonnaise.

PICKLED FIGS.
Place figs in jars with cinnamon bark and cloves. Make a syrup of ½ c sugar and 1 qt. vinegar. Boil and pour into jars. Seal tight.

SPICED FIGS.
1 pt. vinegar	4 lbs. sugar
7 lbs. figs	Cinnamon, cloves

Make a syrup of the vinegar and sugar. Season with 2 parts cinnamon to 1 part cloves, tied in a cloth, and boil 20 minutes. Wash and wipe figs. Drop them in the syrup and boil 10 or 15 minutes, or until tender when pierced with straw. Take out and place in sterilized jars with 2 slices of lemon in each. Boil syrup down and pour in jars.

FIG PRESERVES.
Take equal weight of figs and sugar. Boil sugar with a little water and add figs. Boil, remove and spread on flat plates to cool. Repeat this two or three times and cook until fruit is clear.

Fig   29

DRIED FIGS.

Take figs when the skins begin to crack. Pour over boiling water in which a little soda has been dissolved. Take the figs out and then dip again. Place on trays exposed to the sun for a few days.

FIG JAM,

Peel ripe figs and boil in double boiler until soft. Strain through colander and add ½ as much sugar; boil again and add lemon or pineapple for flavoring.

FIG JELLY.

To 1 doz. figs allow 1 lemon. Peel and slice figs and add unpeeled and sliced lemon. Cook slowly until perfectly soft. Strain through cloth bag and add as much sugar as fruit juice.

FIG PUDDING.

3 large or ½ lb. fresh figs   1 c brown sugar
1 c milk   ¾ c suet
I pt. bread crumbs   1 egg

Chop figs and suet together. Add bread crumbs, sugar, milk, and egg. Boil 4 hrs. in double boiler. Serve with lemon sauce.

FIG CANDY.

2 c sugar   8 figs
2 tbsp butter   ½ c milk

Make a syrup of the sugar, milk, and butter. When it boils, add figs chopped very fine. Beat until creamy. Put in a buttered pan to cool and cut in squares.

FIG PIE.

Allow ½ lb. figs for one pie. Chop figs very fine and cook with 1 c cold water, juice 1 lemon, and little sugar. When soft and smooth, cool and add yolk of egg. Put in pie crust and bake. Remove and cover with a meringue made of the white of egg and 1 tbsp sugar flavored with vanilla. Brown in oven.

FIGS IN JELLY.

1 doz. large figs	4 tbsp sugar
½ c cold water	1 c whipped cream
½ box gelatine	1 tsp vanilla extract

Cover figs with boiling water, and cook slowly until tender. Pour off juice, and boil until reduced to 1 c. Add sugar and gelatine soaked in cold water. Strain, and when cool add vanilla and red coloring. Mold in shallow dish, and when firm cut into dice. Arrange figs in glass dish, and garnish with jelly and whipped cream.

FIGS AND ORANGE JELLY.

½ box gelatine	1 c orange juice
½ c cold water	Juice 1 lemon
½ c sugar	Figs

Take 1 c sliced figs, cover with water, and cook slowly until tender. Strain, and put 1 c juice in saucepan; add sugar and gelatine dissolved in cold water, and lemon and orange juices. When it thickens, add figs. Serve with cream.

CAKES OF FIGS.

Stew slowly ripe peeled figs to smooth pulp. Add little sugar and flavor. Stir constantly till reduced to a thin pulp free of lumps. Pour into pans or molds and dry in the stove, sun, or dryer. When perfectly dry, wrap each cake in oiled paper and keep in a dry place.

FIG SOUFFLE.

Beat 1 c of mashed figs with the whites of 3 eggs and bake in the oven.

FIG FILLING.

10 chopped figs	½ c sugar
1 c water	

Boil together until thick. When cool, spread between and on top of cake.

FIG ICE CREAM.

Mix 1 qt. of cream with 2 c sugar and freeze partially. Then add 2 c mashed figs, and finish freezing.

Fig   31

FIG SHERBET.

1 c water	1 c water-lemon juice
2 c sugar	2 c mashed figs

Boil the water and sugar. Cool and add the water-lemon juice. When partly frozen, add the mashed figs.

FIG FRAPPE.

Take about 1 doz. figs, skin, slice fine, and sweeten. Whip 1 pt. of cream add figs. Pack in freezer without stirring, about 1½ hrs. before serving.

GRAPES

With the exception of the dates, grapes exceed all other fruit in amount of sugar, which varies from 12 to 26 per cent. No sugar is needed when fruit are preserved in grape juice.

CANNED GRAPES.

Wash grapes; put in sterilized jars and steam for 30 minutes. Then fill jars with boiling syrup made of 2 parts water and 1 part sugar. Screw covers on tight.

GRAPE JUICE.

Put clean grapes into porcelain kettle with little water, and cook slowly until seeds and pulp separate. Strain through a cheese cloth, and return juice to kettle. When at boiling point, add ½ as much sugar as juice and dissolve. Boil slowly for about 6 minutes. Fill sterilized bottles and seal at once. For a drink, add lemon juice or slices of lemon and cracked ice.

GRAPE SYRUP.

7 lbs. grapes   2½ oz. tartaric acid

Put grapes into large stone jar and cover with water. Dissolve acid in little water and pour over fruit, mixing thoroughly. Let stand 24 hours. Strain carefully and to each pt. juice add 1¼ pt. sugar, and set aside, stirring occasionally until dissolved, which will take from 12 to 24 hours ; then bottle. It will never ferment. As there is no cooking, the fresh flavor of the grapes is preserved. For drink, pour little into glasses with crushed ice and aerated water.

GRAPE JELLY.

Wash slightly green grapes after taking from stems. Add little water, cover and cook 10 or 15 minutes until soft. Strain juice through cloth and add equal measure sugar. Boil uncovered until it jellies.

GRAPE MARMALADE.

Squeeze off skins of grapes. Cook pulp and skins separately with very little water. When seeds drop out of pulp, strain and add ⅔ c sugar to each c pulp and sweetened skins. For grape butter, cook slowly for a longer time.

SPICED GRAPES.

Cook slowly pulp and skins of grapes separately. Strain pulp; then add to it the skins. To 1 c pulp add ½ c sugar, 2 or 3 cloves, other spices if desired, and 1 tbsp vinegar. Simmer until a rich color.

GRAPE PIE.

Use grape marmalade between 2 crusts, the top slatted.

GRAPE FILLING FOR CAKE.

Mix grape preserve with cream and put between layers of cake just before serving.

GRAPE SALAD.

Remove seeds and skins from grapes, combine with sliced bananas, and serve with mayonnaise.

GRAPE SPONGE.

¼ box gelatine   ¾ c sugar
¼ c cold water   1 c grape juice
Whites 2 eggs   Juice 1 lemon

Soak gelatine in cold water and dissolve by standing dish in hot water. Dissolve sugar in fruit juice and strain the gelatine into the mixture. Set in ice and water and when it thickens add beaten whites of eggs, and beat with egg beater until it is light and spongy. Put into glass dish or shape in mold on ice. Serve with cream or boiled custard made from the yolk of the eggs.

GRAPE JUICE JELLY.

1 box gelatine	1 c cold water
1 pt. grape juice	1½ c sugar
Juice 2 lemons	1½ pts. boiling water

Soak gelatine in cold water 5 minutes. Dissolve in boiling water. Strain and add lemon and grape juices and sugar. Set in ice water, and when it thickens add grapes cut in halves and seeded. Mold on ice and garnish with violets when served.

GRAPES IN JELLY.

½ box gelatine	Grated rind and juice 1 lemon
1 c sugar	½ c powdered sugar
1 c grapes	2 c grape juice

Soak gelatine in grape juice, and dissolve over boiling water. Add sugar and lemon juice. When dissolved, add grated rind and powdered sugar. Set in ice water to thicken; then add seeded and skinned grapes and place on ice.

GRAPE WHIP.

1 pt. cream	¾ c sugar
1½ c grape juice	Whites 2 eggs

Beat the egg whites and add fruit juice mixed with sugar. Add cream and beat. As froth rises, take it off and drain on a sieve. Pour the unwhipped mixture into glasses, and pile with whip on top. Serve cold.

GRAPE JUICE CHARLOTTE RUSSE.

¼ box gelatine	¼ c cold water
¼ c boiling water	1½ c cream
1 c grape juice	Lady fingers
Juice ½ lemon	

Soak gelatine in cold water 5 minutes and dissolve in boiling water. Add grape and lemon juices. Place in ice water, and when it thickens fold in cream. Turn into mold lined with lady fingers. Serve garnished with violets.

GRAPE SHERBET.

3 c water	White 1 egg
1½ c grape juice 1 c sugar	
Juice 1 lemon	

Boil sugar and water, and when cool, add fruit juice. Freeze partly, add egg white and finish freezing.

GRAPE MOUSSE.

1 tsp gelatine	4 tbsp cold water
3 tbsp sugar	½ tsp vanilla extract
1 c grape juice	1 pt. cream

Soak gelatine in cold water, and dissolve over boiling water. Add sugar, vanilla and grape juice. Beat well and fold in Whipped cream. Place in mold or freezer packed in salt and ice for 4 hours.

GRAPE ICE CREAM.

Add 1 pt. grape juice to 1 qt, cream or custard mixture when partly frozen. Use fresh or preserved grapes.

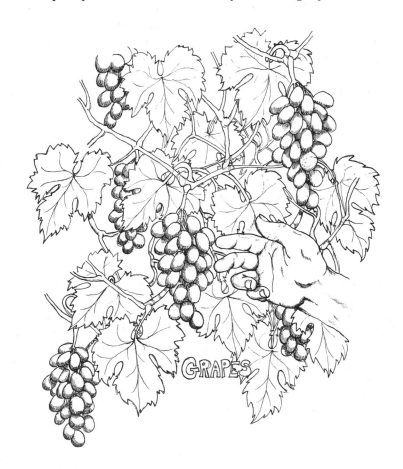

GUAVA

There are several varieties of guavas, the sour and the sweet red, the sweet white, and the small strawberry. As tin discolors guavas, the pulp should be pressed through a net or hair sieve.

Guavas may be dried in an evaporator, to keep for home use when out of season, or exported to be used for jellies and preserves.

GUAVAS SIMPLEST.

Peel, slice, and sprinkle liberally with sugar. Let stand several hours. Serve cold.

GUAVAS SERVED IN SKIN.

1. Take firm, large guavas. Cut off stem ends and with spoon loosen pulp around center. Mix this with powdered sugar. Serve placed on guava leaves arranged on plates.

2. Take pulp out of guava and strain through net, extracting seeds. Mix with little sugar, or sugar and cream, and place in skin again.

STEWED WHOLE GUAVAS.

Take skins from guavas, and pierce with a fork. Drop in boiling syrup made of 2 parts water and 1 part sugar. Cook till tender, but do not break. These are for immediate use.

STEWED GUAVAS WITHOUT SEEDS.

Peel skin from guavas carefully with silver knife. Cut them in halves lengthwise, and take out seeds with a spoon. Place seeds and skins in a saucepan with little water and simmer. Cut halves into long strips. When juice is well cooked from seeds, strain through net, sweeten, return to stove and gently drop the strips into hot juice. Do not stir, but press down syrup and remove scum. Simmer until tender, keeping the form of the strips. If a marmalade is desired, boil until syrup jellies when dropped, and cooled on a saucer. For this add ¾ as much sugar as pulp and juice.

GUAVA SYRUP.

1 pt. peeled and sliced guavas   1 pt. sugar
¾ pt. water

Boil the sugar and water until it strings. Add guava pulp, and cook till translucent. Strain off the syrup. Reheat and pour into sterilized bottles. Seal well. Dilute for use with griddle cakes, in ices, and drink when fresh fruit cannot be had.

GUAVA MARMALADE.

Cut ends off guavas, slice, add little water, and cook until soft. Strain free from seeds in a colander. Add ½ to ¾ as much sugar as pulp, and cook over an asbestos mat, stirring occasionally until it thickens.

SPICED GUAVA MARMALADE.

Take the same as above, adding ½ tsp cinnamon and ¼ tsp cloves to each qt. of fruit pulp.

GUAVA JELLY.

Cut ends off guavas, slice, add little water, and cook until tender. Pour into cloth bag, and let juice drip over night. In morning add equal amount sugar as juice, and boil, stirring until sugar is dissolved. When it boils up hard, test by dropping from a spoon on a saucer. If it jellies when cool, the jelly is done. Pour into sterilized glasses, and seal when cool with melted paraffin. For a tart jelly to serve with meats, use less sugar. In order to have a light colored jelly, make in small quantities and cook quickly. Make remaining pulp into marmalade.

GUAVA JELLY SAUCE.

Add a little boiling water to jelly. Spice and thicken with cornstarch if desired. Serve with cottage pudding.

GUAVA JELLY ROLL.

Make a sponge cake, and bake in large pan. When done, place on a cloth, spread with guava jelly, and roll. Sprinkle with powdered sugar, and roll tight in cloth.

GUAVA MARMALADE COOKIES

1 c sugar	½ c butter
½ c guava marmalade	¼ tsp soda
Cinnamon	

Cream butter and sugar together. Add cinnamon and guava marmalade, the soda dissolved in 1 tbsp boiling water, and flour enough to roll thin.

GUAVA MARMALADE CAKE.

½ tsp cinnamon	1 c guava marmalade
¼ c buttermilk or sour milk	2 c flour
1 c sugar	¾ c butter
½ tsp soda	3 eggs
¼ tsp cloves	

Cream sugar and butter, add beaten eggs, marmalade, spices, soda dissolved in the milk, and flour. Bake ½ hour in hot oven.

GUAVA JUJUBE PASTE.

Make as for jelly, allowing it to cook until it becomes tough. Use in fondant, or cut in squares and wrap in paraffin paper.

GUAVA SAUCE.

1 tbsp butter	1 c guava juice
1 c sugar	1 tbsp flour

Stew guava and pour off juice. Melt butter until it bubbles, add flour, mixing thoroughly. Add guava juice in which sugar has been stirred. Let boil up well, and serve hot with cottage pudding. Guava jelly may be used in place of the juice, melting it with boiling water and omitting the sugar.

GUAVA IN JELLY.

½ box gelatine	½ c cold water
1 c guava pulp	1½ c boiling water
¾ c sugar	

Soak gelatine in the cold water, and dissolve in the boiling water. Strain, and add sugar and guava pulp, which has been pressed through a net. Pour in mold; and put on ice. Serve with cream.

GUAVA SPONGE.

1 c guava pulp	¼ box gelatine
¾ c sugar	¼ c cold water
Whites 2 eggs	

Make as grape sponge and serve with cream.

GUAVA WHEY AND CHEESE.

4 guavas	4 tbsp sugar
1 pt. milk	1 junket tablet

Press guava pulp through net. Crush junket tablet in tbsp cold water. Put milk on stove, and when lukewarm, remove. Add dissolved junket tablet and guava pulp. The whey of the milk will separate from the curd. Pour off whey, and serve iced as a drink. It can be used as lemon whey is used in invalid cookery. Serve curd as guava cheese.

GUAVA WHIP.

4 tbsp sugar	1 c guava pulp
White 1 egg	

Press guava pulp through net. Beat white of egg. Add sugar to guava pulp, and beat into egg white. Serve iced in sherbet glasses. Cream may be added.

GUAVA SOUFFLE.

Powdered sugar	Salt
Whites 4 eggs	1 c guava pulp

Beat egg whites stiffly with pinch of salt. Add guava pulp which has been sweetened with powdered sugar. Turn into buttered pudding dish, or individual molds, in pan of hot water, and bake in slow oven until firm.

GUAVA SHORTCAKE.

Mash 2 qts. strawberry guavas, and mix with sugar. Let stand 2 hours. Make a biscuit dough, and bake in pie tin. When done and still hot, cut edges, and pull apart with forks. Turn crumb sides up, and butter. Place sugared guava on both sides, leaving juice in container. Place one layer on other. Serve sugared guava juice as sauce with shortcake.

GUAVA PIE.

Fill pie shell with guava marmalade and bake. When done

add meringue made of 2 egg whites and 2 tbsp sugar, and brown in the oven. Serve hot or cold.

GUAVA SNOW WITH CUSTARD SAUCE.

2 tbsp cornstarch	½ c boiling water
2 tbsp cold water	¾ c sugar
Whites 2 eggs	1 c guava pulp (preferably white)

Boil sugar and water. Dissolve cornstarch in cold water and add, stirring constantly. When it thickens, take from fire and add guava pulp. Beat this into egg whites which have been previously whipped stiff. Heap in glass serving dish and place on ice. For custard sauce, heat 1 c milk, and 2 tbsp sugar, and beaten yolks of eggs. Do not boil, but steam until it coats a knife blade. Pour this around the snow when it is served.

STEAMED WHIP WITH GUAVA SAUCE.

Make whip of whites 4 eggs, 2 tbsp sugar, and 1 tsp lemon extract. Put into tin mold, and place mold in container of boiling water on stove. Steam 20 minutes. Invert mold on serving dish. Place around it stewed guavas or guava syrup diluted.

GUAVA MOUSSE.

1 pt. cream	1 c guava pulp
1 c powdered sugar	¼ tsp salt

Whip cream, add salt, sugar, and guava pulp. Pack 4 hrs. in equal parts salt and ice.

GUAVA SHERBET.

1 c water	1 c guava pulp
½ tsp gelatine	1 c sugar

Boil water and sugar 20 minutes; add gelatine softened in little cold water, and strain. When cool, add guava pulp and freeze.

GUAVA ICE CREAM.

1 pt. guava pulp	1 pt. cream
1 pt. sugar	

Scald the cream and sugar, and when cool add the guava pulp. Freeze.

MANGO

A great many varieties of this fruit have been introduced but the common sour, or chutney, and Hawaiian sweet mango can still hold their places.

MANGOES SIMPLE.

To eat a mango, pierce the stem end with a fork or nutpick. Hold it in one hand by the fork, and peel with a silver knife in the other hand. Eat from the fork.

Or slice mangoes and eat as peaches.

PRESERVED MANGOES.

Peel skin from mangoes which are a little under ripe. Place in glass jars and cover with syrup made of 2 parts sugar to 4 parts water. Adjust covers loosely. Place jars in steamer and steam until tender, adding more syrup if necessary. Seal tight.

MANGO JELLY.

Peel and slice mangoes. Add to 7 large mangoes 2½ c water and boil for 20 minutes. Strain through double cheese cloth and add gradually ½ c water and wash down juice. Add equal parts sugar to juice, and boil for 10 minutes, or till it jellies when dropped, and cooled on a saucer.

MANGO MARMALADE-1.

Strain the pulp remaining from the jelly through a sieve to get the strings out and add as much sugar as pulp. Boil 20 to 30 minutes, or until thick and glassy.

MANGO MARMALADE-2.

Peel and slice under ripe mangoes ; cover with water and boil until soft. Strain, add ½ as much sugar as mango and boil again, placing an asbestos mat under container. Cook until glassy and thick.

MANGO AND PAPAYA MARMALADE.

Make the same as mango marmalade 2, using ½ as much ripe papaya as mango. A little lemon rind, cut in small bits, may also be added.

SPICED MANGO MARMALADE.

Make as marmalade and add ¼ tsp cloves and ½ tsp cinnamon to 1 qt. fruit pulp.

STEWED MANGO.

Take mangoes when under ripe. Peel skins and cut in slices. Make syrup of 1 part sugar to 2 parts water, and drop slices in. Cook till transparent.

STEWED RIPE MANGO.

Peel and slice ripe mangoes. Add sugar and cinnamon, and simmer slowly till quite dark.

MANGO SALAD.

Slice ripe mangoes and serve on lettuce leaves, with mayonnaise dressing.

MANGO PICKLE.

Make a syrup of equal parts vinegar and sugar. Place a small bag containing 2 parts cinnamon to 1 part cloves in the syrup. Peel and slice under ripe mangoes and drop in the boiling syrup. Boil until clear. Skim out and place in jars. Boil syrup down to original volume and pour into jars.

MANGO CHUTNEY

Take one 14 inch galvanized bucketful of green chutney mangoes; peel and slice them. Or take four lbs. of sliced mangoes.

1 qt. vinegar	1 lb. bleached almonds
1 lb. currants	¼ lb. green ginger
¼ lb. garlic	½ lb. salt
3 lb. brown sugar	2 lb. raisins
2 oz. yellow chili pepper	

Chop all fine except raisins and currants. Cook 4 hours, stirring to prevent burning ; put in bottles and seal. Mix and let stand over night before cooking if possible. Can be made with sweet mangoes, using the juice of 5 or 6 limes or lemons.

MANGO PIE.

Fill a pie shell with stewed mangoes and bake in 2 crusts. A little cinnamon may be added to the mango.

MANGO SHERBET.

Stew under ripe mangoes in little water. Strain and add 1 c sugar to each 1½ pt. mangoes. Cool and freeze.

OHELO

This is the Hawaiian fruit resembling the American blueberry in shape and texture. It is found on the high mountains of East Maui and Hawaii, especially near the Volcano House.

OHELO AND GUAVA JELLY.

2 qts. ohelo juice   ½ c guava jelly
1½ pts. sugar

Make as other jellies. This combination resembles cranberry sauce.

OHELO AND THIMBLEBERRY JELLY.

Cook 1 measure of ohelos and 2 measures of thimbleberries in separate containers until soft. Strain through cloth bags and combine juices, adding ¾ as much sugar as juice. Proceed as with other jellies.

OHELO JAM.

Boil 2 measures ohelos, ½ measure sugar and few slices lemon for 1 hour or more.

OHELO SAUCE.

¼ measure sugar   ¼ measure guava jelly

When boiling, add 2 measures of ohelos and remove from stove before the ohelos have broken, which will be in a very few minutes. The guava jelly may be omitted, but it is an improvement.

OHELO PIE.

Heat ohelos, adding a little sugar and orange marmalade, guava jelly or a slice or two of lemon. Pour off most of the juice. Make pie with two crusts, the top one slatted like the old-fashioned cranberry tart. Cream is an improvement.

OHELO SHORTCAKE.

Make a biscuit dough and bake in a pie tin. Boil 1 c sugar and ½ c water until it strings; then drop in 1 qt. of ohelos and boil until soft. When cake is done, open, butter and fill with part of the berries; remaining berries pour over cake when it is served.

OHIA

The ohia, or mountain apple is found in the market during the season which lasts from July until December. The thin skin contains much of the flavor and should not be removed in using the fruit.

OHIA PRESERVES.

Make a syrup of 2 parts sugar and 4 parts water. Quarter the ohias and place in the boiling syrup. Cook about 45 mins., until the fruit is clear, then fill into sterilized bottles.

OHIA SWEET PICKLES.

Make a syrup of 1 pt. vinegar and 3 lbs. of sugar. Season with 1 part cloves to 2 parts cinnamon, tied in a cloth bag. When it boils, drop fruit in and cook about 30 mins., or until fruit is clear. The ohias may be pickled whole, in halves, or quarters. If cooked whole, pierce fruit with cloves.

OHIA PIE.

Line a pie tin with paste and fill with sliced ohias. Stew over this ½ c sugar, bits of butter, and juice of ½ lemon. Bake about 15 mins; then add upper crust and bake until brown.

OHIA SNOW SHERBET.

1 pt. water	1 pt. fruit juice
1 c sugar	Juice ½ lemon
1 tsp gelatine	

Take ripe ohias which are a deep red color cut into small pieces, and squeeze the juice out through a cloth. Boil the sugar and water for 20 minutes, add softened gelatine, and strain; when cool add the fruit juice and freeze.

ORANGE

Hawaiian oranges are juicier, and contain less of the white, pithy portion under the skin, than the imported ones.

ORANGE DRINK.

Juice 12 oranges   Sugar
1 pineapple   Juice 3 lemons

Cut pineapple in small pieces. Mix with the juice and sweeten to taste.

ORANGE MARMALADE.

Wash oranges. Cut in halves crosswise. Squeeze out juice and seeds. Take out white, pithy part of pulp in the skins. Put the skins and pulp through a meat chopper. Place in saucepan of water on the stove and boil, changing water 3 times, the last time adding orange juice to skins. Measure and add equal measure of sugar. Boil till it becomes glassy, 2 hrs. or longer.

Or take off orange peel in quarters. Boil, changing the water 3 times; then cut into strips and add to the orange pulp, which has been freed from seeds. If sweet oranges are used, add juice of 1 lemon to every 4 oranges. Boil, add equal measure of sugar, and boil again, until thick and glassy. This may take an hour.

ORANGE JELLY.

Juice 1 lemon   ½ c cold water
½ box gelatine   1 c boiling water
1 c sugar   1 pt. orange juice

Soak gelatine in the cold water, and dissolve in the boiling water. Add sugar and stir until dissolved and cooled. Strain and add the orange and lemon juice. Pour into mold or orange skins.

ORANGE AMBROSIA.

Slice 6 oranges and cover with sugar. Grate ½ coconut. Arrange in glass dish in alternate layers of orange and coconut, and heap coconut on top.

ORANGE AND MINT.

Pulp 4 oranges   2 tbsp finely chopped mint
2 tbsp lemon juice   2 tbsp sugar

Mix all together. Place on ice and serve in sherbet glasses as a salad course.

ORANGE STRAWS.

Cut orange skins in strips and boil, changing the water 3 times, until the peel is tender and no longer bitter. Make a syrup of equal parts sugar and water, and cook peel in it 10 minutes. Lift out with a fork, or skimmer, on a plate. When cool, roll in granulated sugar. Place on buttered plate. Roll again in sugar the next day.

ORANGE SPONGE.

1 c sugar   ½ c boiling water
½ box gelatine   1 pt. orange juice
½ c cold water   Whites 4 eggs

Soak gelatine in the cold water 5 minutes and dissolve in the boiling water. Add sugar and stir until dissolved and cooled; then add the orange juice. Place in a dish of ice water until it thickens; then beat in the whites of eggs which have been whipped to a stiff froth. Beat with a Dover egg beater until smooth. Turn into a mold and place on ice.

ORANGE CUSTARD

5 or 6 oranges   1 pt. milk
1 c sugar   3 eggs
1 tbsp cornstarch

Peel oranges and cut into small pieces, taking out the white pulp. Place in a pudding dish covered with 1 c sugar. Make a custard of the milk and yolks of eggs, adding 2 tbsp sugar and the cornstarch dissolved in a little cold milk. When cool, pour this over the oranges. With the egg whites and 2 tbsp powdered sugar make a meringue and spread on pudding. Place in oven to brown or serve without browning.

ORANGE CHARLOTTE.

½ box gelatine   1 c cold water
1 c sugar   1 c orange juice and pulp
Whites 3 eggs   Juice ½ lemon
2 c whipped cream

Soak gelatine in water and dissolve over boiling water. Strain and add sugar, lemon juice, orange juice and pulp. Place in ice water, and when it thickens, stir until frothy; then add whites of eggs stiffly beaten, and fold in cream. Line a mold with orange sections, and turn in mixture. Set on ice.

ORANGE WHIP.

Juice 1 orange	1 pt. cream
2 tbsp sugar	2 tbsp grated coconut
1 egg	Grated rind of orange

Mix sugar and coconut with the orange juice and grated rind. Whip the cream, and beat the mixture into it. Place on ice and serve cold.

ORANGE MARSHMALLOW.

1 lb. marshmallows	Juice 2 oranges
Juice 2 China oranges	Sugar

Cut each marshmallow into 4 pieces and squeeze over them the strained juice of the fruit. Add ¼ c or more of sugar. Mix all in bowl and set on ice. Serve in sherbet glasses with a maraschino cherry on top each glass. Pink marshmallows may be used.

ORANGE SALAD.

Slice 4 peeled oranges. Mix with 1 tbsp lemon juice and 3 tbsp oil. Arrange in heap on lettuce leaves. In the center place 1 c chopped nuts mixed with 1 tbsp lemon juice.

ORANGE PUDDING FOR TWO.

1 orange	1 egg
1 c milk	2 tbsp sugar

Skin and seed orange. Cut in small pieces, and put in a baking dish. Beat the yolk of egg with sugar. Add milk and pour over orange. Bake pudding and add meringue made of the egg white and 1 tbsp sugar.

ORANGE TART.

Beat together juice of 2 oranges, grated peel of 1, ¾ c sugar, 1 tbsp butter. Add 1 tsp cornstarch dissolved in juice of 1 lemon. Bake in tart shells.

ORANGE PIE.

Cut oranges into small pieces. Place in pie crust and sweeten well. Add seeded raisins, sprinkle with flour and cover with upper crust. Bake in hot oven ½ hour.

ORANGE CAKE.

½ c butter	1½ c flour
3 eggs	1 c sugar
½ c orange juice	½ grated rind of orange
1½ tsp baking powder	¼ tsp salt

Cream butter and sugar, and add beaten yolks of eggs, salt, and grated rind of orange. Beat in orange juice and flour which has been sifted with the baking powder. Fold in stiffly whipped whites of eggs. Bake in loaf, and when cool spread with orange frosting.

ORANGE FROSTING.

Mix juice and grated peel of ½ orange with powdered sugar to make stiff enough to spread on cake.

ORANGE FILLING.

Juice 1 orange	1½ c sugar
¼ grated rind	1 tsp butter
Yolks 2 eggs	

Beat eggs, add sugar, grated rind, butter, and juice of orange. Cook over hot water, stirring until it thickens. Spread between layers of cake.

ORANGE SHERBET.

1 c sugar	1 c orange juice
1 pt. water	1 tsp gelatine

Boil water and sugar, and add gelatine soaked in little water 5 minutes. When cool, add orange juice. Strain and freeze.

PAPAYA

The milky juice of the papaya contains a digestible principle similar to pepsin. Its leaves have the property of making tough meat tender when wrapped in them over night.

PAPAYA PUNCH.

Make a strong lemonade of 6 lemons, add 1 medium sized papaya cubed, and 5 or 6 bottles of ginger ale, according to amount desired.

STEWED PAPAYA-1.

½ c sugar   Juice 2 lemons
2 c diced papaya   ¼ c water

Cut papaya in dice, and stew with sugar, water, and lemon juice ½ hr. Serve in sherbet glasses as a first course for luncheon, or a dessert. Can use 4 China oranges in place of lemons.

STEWED PAPAYA-2.

Cook in the same manner as 1, with ¼ c sugar and only enough water to keep from burning. Serve as a vegetable.

STEWED GREEN PAPAYA.

Take a green papaya which has no signs of yellow. Skin and cut in large pieces, and boil. Serve as a vegetable. It may be mashed like squash.

BAKED PAPAYA.

Cut papaya in halves lengthwise. Add a little sugar and China orange, lime or lemon juice; or a little cinnamon in place of the juice. Bake 20 minutes, and serve immediately on taking from the oven. This is a vegetable.

PAPAYA AND CHINA ORANGE MARMALADE
(See China Orange.)

PAPAYA AND TAMARIND.
(See Tamarind.)

PAPAYA PICKLE.
Make syrup of 1 measured sugar and ½ measure vinegar. Add a few whole cloves and pepper corns and 2 measures of half ripe papaya cut into small pieces. Boil until tender.

PAPAYA AND GINGER.
Take a syrup of 1 measure sugar, ½ measure water some finely sliced dried ginger, and a few slices of lemon. Add 2 measures half ripe papaya sliced lengthwise, which has been previously simmered in water until clear, but not broken.

PAPAYA AND BANANA.
Put both fruits through round potato cutter and serve with lemon juice.

PAPAYA COCKTAIL.
Cut papaya in dice and serve in glasses with cocktail sauce and chipped ice.

Or serve with China orange, lemon or lime juice, and a little sugar in same manner.

PAPAYA SALAD-1
On a strip of peeled papaya lay small bits of pomelo and orange. Serve with mayonnaise on separate plates, and garnish each with one or two nasturtium flowers and leaves.

PAPAYA SALAD-2.
Cut papaya in cubes, and add 8 small Chinese onions and 5 pieces green celery chopped fine. Serve with boiled dressing.

PAPAYA WHIP.
To 1½ c papaya pulp add juice of 1 lemon, ½ c sugar and beat into 2 stiffly whipped whites of eggs.

PAPAYA JELLY.

½ box gelatine	½ c sugar
½ c cold water	1 c boiling water
Juice 1 lemon	1 c papaya pulp

Soak the gelatine in the cold water 5 minutes. Dissolve

the sugar in the boiling water; add the gelatine, and strain. When cool, add the papaya and lemon juice. Place on ice to harden.

PAPAYA BISCUIT.

Use a baking powder biscuit recipe. Cut in rounds with a cutter and place rounds in muffin tins. On top of each round put a piece of papaya flavored with sugar and lemon juice. Bake and serve hot.

PAPAYA PIE.

2 eggs	1 c sugar
1 c papaya pulp	Juice ½ lemon
½ c butter	

Make a bottom pie crust and bake. Cream butter and sugar. Add beaten eggs, lemon juice, and papaya. Pour into pie crust and bake. Make a meringue of whites of eggs and 2 tbsp sugar. Place on pie and brown in oven.

PAPAYA SHERBET.

1 pt. water	1 pt. papaya pulp
1 c sugar	Juice ½ lemon
½ tsp gelatine	

Boil water and sugar 20 min., add gelatine softened. Strain and when cold add fruit and freeze.

PAPAYA SHERBET WITH MILK.

Juice of 3 lemons	Pulp good sized papaya
1 qt. milk	1 pt. sugar

Soak sugar and lemon juice 2 or 3 hrs. When ready to freeze, add sugar and lemon juice to papaya and milk. It will curdle before freezing.

PINEAPPLE

The pineapple is the only fruit known to contain a vegetable pepsin which aids digestion. In order to get the full benefit of this digestive principle, the fruit should be eaten fresh and without sugar. This vegetable pepsin dissolves albuminous substances, and hence the juice should be scalded before combining with milk, eggs, or gelatine. Tough meat placed between slices of pineapple over night will become tender. Pineapples do not sweeten if picked green.

TO SERVE FRESH PINEAPPLE.

Slice or grate pineapple; sprinkle with sugar and set in the ice box for several hours before serving.

PINEAPPLE JUICE.

Press out juice from pineapple and strain through flannel. To each measure of juice add ½ measure of sugar. Bring to boiling point, or stand in rapidly boiling water for 1 hour. Bottle while hot in sterilized containers. Use with lemon and water as a cold drink, or in punches, puddings and sauces. It is good in hot or cold tea, with or without lemon.

PINEAPPLE MARMALADE.

Allow 1 measure sugar to 2 measures grated pineapple. Mix well and let stand over night. Next morning cook slowly 2 hours or more.

PINEAPPLE PICKLE.

Take 1 measure vinegar, 3 measures brown sugar, a few whole cloves and pepper corns. Boil together. Add 4 measures pineapple cut into small pieces, and boil until fruit is golden yellow. If the syrup is not rich, boil down before pouring it over the fruit. If preferred, use 1 tsp cinnamon to ½ tsp cloves tied in cloth for spicing. The pickles are best made in pineapple vinegar.

PINEAPPLE VINEGAR.

Press out juice from fresh pineapple and strain through double cheese cloth. Place in jar and leave to ferment 2 or 3 weeks. It improves with age.

PINEAPPLE IN OHELO JUICE.

Boil slowly together 1 measure of minced pineapple and 1 measure of ohelo juice until the pineapple is tender. Then add ½ measure of sugar and boil slowly for 2 or 3 hours, or until the fruit is the color of quince preserve.

PINEAPPLE CHIPS.

Cut pineapple into thin strips, lay on china dishes, and cover with granulated sugar. Keep strips separate. Dry in the sun or hot air closet. Sprinkle with sugar and turn every day, pouring off the syrup. When the chips are dry or crystallized, pack in jars or tin boxes with oiled paper between.

CANDIED PINEAPPLE.

To 1 measure of thickly sliced and cored pineapple add 1 measure of sugar. Boil together until the fruit is of a rich golden color. Take from the syrup, spread on dishes in the sun, turning every day until dry.

PINEAPPLE AND OHELO JELLY.

Boil pineapple and ohelos separately. Strain through jelly bag. To 1 measure pineapple juice and 1 of ohelo juice add 1 measure white sugar. Cook the same as other jellies. The ohelos bring out the pineapple flavor.

PINEAPPLE AND GUAVA JELLY.

Cut ends off guavas, add a little water and cook until soft. Peel equal amount of pineapple. Cut in small pieces and cook until soft. Strain both juices through a bag. Add equal amount of sugar and boil again, stirring until the sugar is dissolved. Skim off scum, and when it boils up hard, test by dropping on a saucer. If it jellies when cool, it is done. Pour into sterilized bottles and seal.

PINEAPPLE CHUTNEY.

2 oz. pepper	1½ lbs. brown sugar
1 oz. garlic	1½ pts. vinegar
3 lbs. pineapple	½ lb. raisins
2 tbsp ginger	½ lb. almonds
1 tbsp salt	

Cook pineapple, vinegar, and salt. Chop nuts, raisins, garlic, and pepper fine, and add to pineapple. As the pineapple is juicy, it requires boiling down.

PINEAPPLE SALAD.

Pare and cut in dice 1 pineapple. Wash and cut fine ½ as much celery. Whip ½ c cream or white of 1 egg into mayonnaise, and mix part with the pineapple and celery. Place this on lettuce leaves. On top, heap remaining mayonnaise and garnish with celery tips.

PINEAPPLE AND AVOCADO SALAD.
(See Avocado.)

PINEAPPLE MINCE MEAT-1.

Mix together 1 measure minced cooked pineapple, ½ measure raisins and currants, some thinly sliced citron or orange peel, or both. Add powdered cinnamon, cloves, and nutmeg to taste and syrup from sweet pickles or sweetened vinegar. Before making the pies, add chopped bananas or apples.

PINEAPPLE MINCE MEAT-2

1 pt. chopped meat	3 c brown sugar
3 pts. crushed pineapple	1½ c molasses
with juice (canned or fresh)	Butter size of egg
1 extra c pineapple juice	2 tsp cinnamon
1½ c raisins	2 tsp allspice
1 c currants	1 tsp cloves
1 tbsp salt	½ tsp pepper

Simmer all together for an hour, stirring often. Add 1 tbsp cornstarch dissolved in a little cold water, and boil again. Then add 1 c wine or brandy, and seal in glass jars. When making the pies, heat the meat, and add 2 tbsp brandy to each pie.

PINEAPPLE PIE.

1 grated pineapple	2 tbsp flour
1 c water	2 c sugar
2 eggs	Salt

Mix all together, and bake in 2 crusts. This makes 2 large or 3 small pies.

PINEAPPLE AND GUAVA PUDDING.

Boil 2 c grated pineapple and ½ c guava jelly very slowly for 20 minutes. Add 1 tbsp cornstarch dissolved in a little cold water, and stir until clear. When ready to serve, whip cream and mix fruit with it.

PINEAPPLE SHORTCAKE.

Make the same as pineapple and guava pudding without cornstarch. Pour between and on top of rich biscuit crust, and serve at once.

PINEAPPLE SAUCE.

To 5 measures grated pineapple add 1 measure white sugar. Boil slowly till clear and seal.

PINEAPPLE JELLY.

1 tbsp gelatine	1 c grated pineapple
½ c boiling water	2 tbsp sugar
Juice ½ lemon or 1 tsp lemon extract	

Soak gelatine in ¼ c cold water and dissolve in the boiling water. Add sugar and lemon juice. Strain, and when cool set on ice for ½ hr. When it thickens, beat in the grated pineapple that has been previously brought to boiling point.

PINEAPPLE CREAM.

2 tbsp gelatine	2 c pineapple juice
½ c cold water	1 c sugar
1 pt. cream	

Soak gelatine in cold water 5 minutes. Dissolve by setting container in boiling water. Heat pineapple juice to boiling point and add sugar. Combine, and let cool in ice water. When it thickens, beat in cream, previously scalded. Place on ice.

PINEAPPLE SPONGE.

1 grated pineapple	⅔ c sugar
½ box gelatine	½ c cold water
Whites 2 eggs	Lemon juice

Let pineapple simmer on stove 10 minutes. Add sugar and lemon juice. Soak gelatine in the cold water and dissolve in pineapple. Strain into mold, let cool and set on ice for ½ hour. Then add stiffly beaten whites of eggs, and beat well. Set on ice until served.

PINEAPPLE JARDINIÈRE.

Cut top from pineapple and reserve for cover. Scoop out the inside and combine with 2 bananas, 2 oranges, and the seedless pulp of 1 c grapes. Add sugar and serve cold in the pineapple rind.

PINEAPPLE TAPIOCA PUDDING-1

3 tbsp instant tapioca	4 tbsp sugar
1 c cold water	2 c fresh pineapple juice
1 tbsp butter	2 eggs

Soak the tapioca in the cold water, add ⅛ tsp salt and sugar. Boil until clear, adding as little water as possible. Beat the yolks of eggs and stir them and the butter into the mixture. When smooth, take from the fire, add pineapple and ½ tsp lemon extract. Turn into a baking dish. Beat the egg whites with 2 tbsp sugar and put on top. Brown 2 or 3 minutes in very hot oven.

PINEAPPLE TAPIOCA PUDDING-2.

½ c minute tapioca	1 small pineapple
3 c hot water	1½ c sugar
2 eggs	

Boil tapioca in hot water until clear. Chop pineapple fine and add to sugar; then stir into tapioca and let boil up. Beat this into the stiffly whipped whites of eggs. Serve cold with a custard made from the yolks of the eggs.

PINEAPPLE AND COCONUT.

Cut fresh pineapple in small triangles, and place in a dish with a mound of grated coconut and sugar on top. Serve as a dessert.

PINEAPPLE SHERBET-1.

1 pt. scalding pineapple juice   ½ tsp gelatine
1 pt. water   Juice 1 lemon
1 pt. sugar

Soak the gelatine in ¼ c cold water for 10 minutes. Add 1 c boiling water, and stir until dissolved. Add sugar and ¾ c cold water; then pineapple juice which has been strained and measured after straining. Freeze. Either canned or fresh pineapple may be used.

PINEAPPLE SHERBET-2.

2 large pineapples   1 pt. water
2 c sugar   Juice 4 lemons
whites 4 eggs

Grate pineapple to extract the juice, strain and add juice of lemons. Add sugar to the whipped whites of eggs. Combine and freeze at once.

PINEAPPLE ICE CREAM.

Scald 1 pt. of pineapple juice. Add juice ½ lemon, and sweeten to taste. Add this to I qt. of any plain cream when partly frozen.

PINEAPPLE MOUSSE.

1 c pineapple syrup   1 tsp gelatine
1 pt. cream   2 tbsp cold water
1 c powdered sugar   ¼ tsp salt

Heat 1 can pineapple and drain juice. Soak gelatine in the cold water and dissolve in the hot syrup. Add salt, lemon juice, and sugar. Strain and cool. Place in ice water and when it thickens fold in whipped cream. Freeze in freezer or mold, packed in equal parts salt and ice for 4 hours.

POHA
The poha is the Hawaiian gooseberry.

POHA JAM.
Stew pohas with a little water until soft. Measure and add equal measure of sugar. Cook until glassy. An asbestos mat under the saucepan will prevent burning.

POHA JELLY.
Cook pohas with a little water until the juice comes well out. Strain through a cloth bag, and add equal measure of sugar. Cook until it jellies when dropped on a saucer and cooled.

Or serve syrup before it has jellied with plain ice cream.

POHA PIE-1.
Fill a pie crust with stewed pohas, and add 1 tbsp butter. Cover with upper crust and bake 30 minutes in hot oven.

POHA PIE-2.
Line a pie dish with pastry, and bake. When done, put in a layer of poha jam, then a custard made of 1 c milk, 1 tsp butter, 1 tsp cornstarch dissolved in a little cold milk, the yolks of 2 eggs, and vanilla flavoring. Spread the sweetened whipped egg whites on top, and brown slightly in hot oven. Serve cold.

POHA JAM COOKIES.
Use a good, but not too rich, cooky recipe. Roll out and cut into strips 4 inches wide. Spread almost half the width with poha jam and fold over the other half, pinching the edges together. Then cut in 3 inch lengths and bake.

POHA JAM SURPRISES.
Make soft, thin baking powder biscuit rounds about the size of saucers. Put a little poha jam on one side, and turn over the other, wetting the edges and pinching together. Bake quickly and serve warm.

POHA SHORTCAKE.

Boil pohas in almost no water for a few moments. There should be very few broken. Pour off most of the juice. Sweeten, and thicken the remainder with a little cornstarch. Serve with biscuit rounds for shortcake or bake between 2 pie crusts.

POMELO AND SHADDOCK

The pomelo is the same fruit as the grapefruit, but the name pomelo is now given the preference. The shaddock is the large, coarse form of the same species.

TO SERVE POMELO.

1. Cut fruit in halves. Separate pulp from skin with silver knife. Cut out white pulp in center, and free juice in lobes by running knife through them. Add sugar and place on ice until served. A maraschino cherry may be placed in center of each half.

2. Cut in halves. Take all pulp out. Mix with sugar and place on ice 1 or 2 hours before serving. Serve for breakfast or first course for luncheon.

POMELO ASPIC SALAD.

Cut and divide pomelo into sections. Lay these on a flat dish and sprinkle with chopped nuts. Take ½ box gelatine and soak in ½ c cold water 5 minutes. Add 1½ c boiling water. When dissolved, strain and season with salt, 1 tbsp lemon juice and paprika. Pour this over pomelo and nuts and when cool set on ice to harden. Serve cut in blocks on lettuce leaves with mayonnaise.

POMELO SALAD.

Cut pomelo into halves. Take out pulp and divide into sections. Mix with chopped nuts, and replace in half skins. Serve with mayonnaise.

POMELO JELLY.

½ box gelatine   2 c pomelo juice and pulp
½ c cold water   Juice 1 lemon
1 c boiling water   1 c or less sugar

Soak gelatine in cold water and dissolve in boiling water. Add sugar, lemon juice, and pomelo juice and pulp. Let partly set before turning into mold, in order to avoid settling of pulp. May be served in baskets made of pomelo skins.

POMELO MARMALADE.

To 2 pomelos allow 2 lemons. Cut pomelos in halves and remove centers and seeds. Remove seeds from lemons and grind all in a meat chopper. Cover with boiling water and let stand over night. Cook till tender, then add 2 c sugar and cook until it thickens and becomes glassy.

Or take off peel of pomelos in quarters and boil, changing the water 3 times; then scrape out the white, pithy portion, and cut skin in strips. Free pulp from seeds and cut in small pieces. Add this to the skins and boil until thick and glassy.

CANDIED POMELO PEEL.

Slice, peel and cover with water, adding 1 tbsp salt to 1 qt. water. Leave over night. In morning drain and place peel in saucepan, a little more than covering it with fresh water, and boil gently for 4 hours. Drain, add cold water as before, and boil till tender. Again drain, weigh peel, and add equal weight sugar and 1 c water to each pound. Boil gently 1 hour with saucepan covered. Remove cover and let simmer till syrup is all absorbed, taking care not to burn. Spread on platter and leave 24 hours, or until it seems fairly dry; then roll each piece in fine granulated sugar, shaking off the sugar to prevent caking. Let stand a few hours to dry; then place in covered jar.

TO SERVE SHADDOCK.

1. Extract pulp from shaddock and add orange juice and a little sugar. Serve iced in sherbet glasses.

2. Extract pulp from shaddock and serve with seeded dried dates or figs which give the required sweetness.

SHADDOCK SALAD.

Extract pulp from shaddock, ice and serve on lettuce leaves with mayonnaise.

ROSELLE

The seed of the roselle is covered with a red calyx which is similar to the cranberry when cooked. It can be cooked, whole, but makes a less stringy substance when the seed is removed from the calyx.

ROSELLE DRINK.

Stew the calyses of the roselle with a little water until soft. Strain the juice and add sugar and water to taste.

STEWED ROSELLE.

Stew the calyxes of the roselle with a little water until soft then add sugar to taste.

ROSELLE JELLY.

Cook the calyxes of the roselle, until soft, in enough water to keep from burning. Strain the juice through cloth bag and add equal measure of sugar, stirring until sugar is dissolved. Skim off all scum. When it boils up hard, test by dropping from a spoon on a saucer. It is done if it jellies when cool.

ROSELLE MARMALADE-1.

Take the pulp left from the jelly. Add a little water if too thick and an equal measure, or more of sugar, if a solid marmalade is desired. Cook over an asbestos mat until it thickens.

ROSELLE MARMALADE-2.

Cook 3 lbs. roselle calyxes in 2 c water for 1 hr. Measure and add 1½. c sugar to each c fruit. Stir or cook over asbestos mat until thick.

ROSELLE SAUCE.

1 tbsp cornstarch	1 c sugar
1 c roselle calyxes	1 tbsp butter

Look the calyxes with a little water until soft. Strain the juice through a colander. Melt the butter, add the cornstarch stirring. When it thickens, add sugar and juice and let boil up. Serve with cottage pudding.

ROSELLE GELATINE.

½ box gelatine	1 c sugar
2 c cold water	Juice ½ lemon
1 c boiling water	1 c roselle calxes

Soak gelatine in cold water 5 min. Cook roselles in boiling water until soft, then strain through sieve. Place juice on fire, add sugar, gelatine, and lemon juice. Turn into a mold and place on ice. Serve with cream.

ROSELLE PIE.

Line a pie dish with pastry and fill with stewed and sweetened roselle calyxes. Cover with upper crust and bake. Papaya pulp may be used with the roselle in pie.

ROSELLE SHERBET.

| 1 qt. water | 1½ c roselle juice |
| 2 c sugar | 1 tsp gelatine |

Boil water and sugar 20 min.; add softened gelatine. Strain, and when cool, add roselle juice taken from stewed roselle calyxes.

TAMARIND
The tamarind contains a laxative and cooling quality which makes it of value in cases of illness.

TAMARIND DRINK,
1. Shell tamarind, and cover with water. Let soak several hours; then take the water from tamarinds, dilute with fresh water, and sweeten to taste.

2. To make drink quickly. Place 10 or 12 shelled tamarinds in a 2 qt. pitcher filled with water. Stir well, and add sugar to taste.

TAMARINDS PRESERVED.
1. Shell tamarinds. Place layer in jar, and cover with sugar. Repeat until jar is filled; then seal. Use candy boxes in same way, placing paraffin paper between layers of tamarinds.

2. Shell tamarinds and press tightly together in form of ball. Cover with cloth. An easy way to carry them in traveling, in which they will keep for years.

3. Make syrup of equal measures sugar and water. Boil tamarinds in this, and pour into jars to be kept for drinks.

TAMARIND AND PAPAYA.
Cook 1 c of shelled tamarinds in a little water until soft. Strain pulp through colander. Add to tamarind pulp, 1 qt. papaya pulp and 1 c sugar. Cook 30 or 45 min.

TAMARIND AND BANANA.
Make the same as above, using banana in place of the papaya.

TAMARIND CHUTNEY.
½ lb. tamarinds	½ lb. dates
½ lb. green ginger	½ lb. raisins
½ lb. onions	¼ lb. chili peppers
4 tbsp brown sugar	2 tbsp salt

Pound all with vinegar, and rub through a sieve. Bottle and seal.

TARO

The taro, or kalo, though not a fruit, is a typical Hawaiian vegetable. It is very acrid, and stings the mouth if under cooked.

BOILED TARO.

Cut skin from taro, divide into 2 inch sections, and cover with salted water in a saucepan. Boil 40 minutes to 1 hour until tender. The skin may be left on, if it is well cleaned, and peeled after boiling. Serve with butter.

TARO AND STEW.

Cook taro with a stew in place of potatoes.

TARO PART BOILED AND BAKED.

Cut skin from taro and boil in salted water 20 minutes. Take from water and place on pan in hot oven for ½ hour Serve with butter and salt.

BAKED TARO.

Cut skin from taro, and if large, cut taro in half. Put in oven and bake for 1 hour. Break with the hands into individual pieces and serve with butter and salt.

FRIED TARO.

Cut boiled taro into thin pieces. Sprinkle with salt, and fry in butter or lard.

BAKED TARO CAKES.

Mash hot boiled taro with a potato masher until free from lumps. Form into small cakes with the hands, using as little water as possible. Place in buttered pan, and bake ½ hour in hot oven.

FRIED TARO CAKES.

Make same as baked taro cakes, and fry in butter, or lard, or in deep fat. Drain and sprinkle with salt. For breakfast taro cakes, boil and mash taro night before. In morning moisten with water, and shape.

PAIAI.

Bake taro and pound until free from lumps. Press and pack it. This will keep for months.

POI.

Dilute paiai with water and expose to fermentation. Eat with salt fish, or as a porridge with sugar and milk. Also use to thicken gravy instead of flour.

POI COCKTAIL.

To a glass of milk add 2 tbsp poi. Stir this into the milk and sweeten if desired. This is served to invalids.

LUAU.

Take the young, tender leaves of the taro plant, and strip stems as with beans. Boil, changing water 3 times and adding a pinch of soda. Cook until tender, 45 minutes in all. After last water, a little milk may be added with salt and pepper. Mix well before serving. Cut hard boiled egg in slices and lay on top.

LUAU AND CHICKEN.

Boil 3 bundles of luau as above. Stew a chicken, and when done, pour off the gravy and mix chicken with the luau. Grate 2 coconuts and pour over the chicken as a gravy. Squid and luau is made in the same way.

LUAU SOUP.

Cook luau as above, using 1 bundle. Boil 1 qt. milk. Add 2 tbsp butter and luau which has been pressed through a sieve. Stir and season with salt and pepper.

TARO FLOWERS.

Take out pistils from flowers and strip strings from stems. Boil flowers and stems in water with pinch of soda, changing water twice. Cook 45 minutes in all. Serve as a vegetable. These can be bought in small bundles at the market.

TARO LEAF STALKS.

Take stalks of taro plant which come on taro above the tuber. Scrape them and slice. Place in cold water for ½ hour, then put into salted boiling water with pinch of soda, and cook until tender. Pour off water. Season with salt and pepper, and serve on buttered toast with a little white sauce.

WATERMELON

Though the watermelon lacks food nutriment, still it is considered one of nature's purifiers.

WATERMELON PRESERVED RIND.

7 lbs. rind   2 lemons
4 lbs. sugar   4 c water

Prepare as for watermelon pickle. Cook until tender; then add sugar, and when thick, sliced and seeded lemons. Cook a few minutes.

WATERMELON PICKLE.

To 12 lbs. rind use 7 pts. vinegar and 1 tsp alum to 1 qt. water. Pare off green part of rind of melon. Cut white part into strips and pour over boiling water in which alum has been dissolved. Leave over night. Drain and soak several hours in fresh water. Cook until tender; then make syrup of equal parts sugar and vinegar seasoned with 1 tsp cinnamon to ½ tsp cloves tied in a bag. Drop melon in, and cook until tender. Skim out, and place in jars. Boil syrup down, and pour over melon.

WATERMELON SHERBET.

Take all red pulp from melon. Add lemon juice, and sweeten to taste. Freeze. When partly frozen, add the whipped white of 1 egg to each qt. of sherbet and finish freezing. It may be served in melon skin.

COMBINED FRUIT

Many combinations of fruit may be served, according to the season and what one has on hand. In combining fruit, cut all in small pieces and mix gently with a spoon.

Or save all the juice in cutting. Place the firmer fruit at bottom of serving dish, then alternate layers, and on top pour the combined juices saved from cutting. It may be served in individual sherbet glasses. Ice before serving.

FRUIT COMBINED-1

Pile minced bananas on a cored slice of slightly divided pineapple, and pour over them a tbsp sweetened orange juice.

FRUIT COMBINED-2.

Combine ½ pineapple chopped fine, 6 sliced bananas, 6 sliced oranges, 1 pt. strawberries, 1 lemon thinly sliced, and sugar to taste.

Or combine the pineapple, bananas, and oranges with ½ oz. grated coconut.

FRUIT COMBINED-3.

Combine 1 banana, 1 orange, ½ c pineapple, ½ c walnuts 12 chopped marshmallows. Just before serving beat the mixture through 1 c whipped, sweetened, and iced cream. Serve at once in sherbet glasses.

TROPICAL PUDDING.

½ c pearl tapioca	1 orange sliced
1½ c sugar	1 banana sliced
¼ tsp salt	2 slices pineapple cubed

Soak the tapioca over night. In morning add to it 1 pt. boiling water, and boil in double boiler until clear. Cool slightly, add all the fruit, mix together, and place on the ice. Serve with cream.

FRUIT MARMALADE.

Take 4 parts papaya pulp, 1 part pineapple, 1 part orange juice and skins sliced fine. Cook all together until soft, then add equal measure or more of sugar. For a solid marmalade more sugar is required.

FRUIT CREAM.

Wash 4 bananas. Add 2 oranges, pulp and juice, and 1 tsp lemon juice, and mix with ⅔ c powdered sugar. Soak ¼ box gelatine in ¼ c cold water 5 minutes and dissolve over boiling water. Add this to the fruit mixture and cool in ice water. When it thickens, fold in 2 c whipped cream. Place on ice and serve cold.

FRUIT JELLY.

Slice 1 orange, 2 bananas, and 4 figs. Seed 6 dates and ½ c grapes, and chop 16 almonds. Soak ½ box gelatine in ½ c cold water 5 minutes. Make a syrup of 1½ c boiling water and 2 c sugar, and dissolve the gelatine in it. Strain and place in ice water. When it thickens, stir in the fruit. Place on ice, and serve with whipped cream.

FRUIT SALAD-1.

Take fresh pineapple, papaya, and celery. Cut in small pieces, and add seeded grapes. Serve on lettuce leaves with mayonnaise dressing.

FRUIT SALAD-2.

Fill half a pomelo skin with minced bananas, pomelo, apples, and a very little pineapple. Put sliced or chopped walnuts on top, and cover with mayonnaise.

FRUIT SALAD-3.

Slice 2 bananas, 2 apples, 1 small pineapple, and 1 papaya. Mix the fruits and serve on lettuce leaves. Cover fruit with whipped cream.

FRUIT PUNCH-1.

Take juice of 3 lemons, 2 oranges, and 1 pomelo. Add 1 qt. water (part mineral water), and large slices of pineapple. Sweeten and serve iced.

FRUIT PUNCH-2.

1 c grenadine	1 c orange juice
½ c lemon juice	1 c pineapple juice
1 c strong tea	1 bottle maraschino cherries
Sugar or sugar syrup.	

Combine all, sweeten to taste, and add mineral water, if desired.

FRUIT PUNCH-3

6 oranges	1 pineapple
6 lemons	sugar
1 papaya	ginger ale

Cube pineapple and papaya. Add juices of oranges and lemons. Sweeten to taste and add 5 or 6 bottles of ginger ale when ready to serve.

POI LUNCHEON COOKED OVER ORDINARY FIRE

TABLE ARRANGEMENT.

Cover the table cloth with ferns. At each place put a bowl of poi, salt dish, or shell containing kukui nut, or mamona, a small side plate with native salt, a few cooked shrimps, native onions, radishes, and chili peppers.

One can buy the cooked kukui nut, or mamona at the market. Crack the shells open and take out the kernel. Pound this with Hawaiian salt in a wooden dish, or mortar. The Kauai red salt can be bought at the market by asking for paa-kai ulaula o kauai.

FIRST COURSE.

Salmon lomilomi. A salmon belly, 5 cents tomatoes, 5 cents onions. Soak salmon over night, pick to pieces, removing white membranes. Slice onions and tomatoes, remove skins, add 3 or 4 whole chili peppers, mix all thoroughly and place on ice. Serve in large glass dish with lump of ice.

SECOND COURSE.

Mullet, or kumu, small ones for individual service if possible. Clean fish, sprinkle with coarse salt, tie up in ti leaves, and broil.

THIRD COURSE.

Chicken, luau, sweet potatoes. Old fowl may be used. Cut chicken in pieces, scald luau in three waters and scrub potatoes well. Take agate bowl that will fit your steamer, and line with banana leaves. Pack tight into it the chicken, scalded luau, and fill up any vacancies with sweet potatoes. Scatter in coarse salt, cover all with ti leaves, about four thicknesses of banana leaves, folded cloth, and then cover of the steamer. Weight this down with stones or wood, so that it will be impossible for any steam to escape, and do not open until ready to serve. Cook 5½ to 6 hours. A fireless cooker can be used for this.

FOURTH COURSE.

This may be an Hawaiian pudding, such as *papaiee* (page 17), *koele palau,* or *haupai* (page 21). Or serve ice cream, cake, and coffee after the guests have left the table.

Books by the
PETROGLYPH PRESS

ABOUT HAWAII'S VOLCANOES *by L. R. McBride*
A CONCISE HISTORY OF THE HAWAIIAN ISLANDS
by Phil K. Barnes
HILO LEGENDS *by Frances Reed*
HOW TO USE HAWAIIAN FRUIT *by Agnes Alexander*
JOYS OF HAWAIIAN COOKING
by Martin & Judy Beeman
THE KAHUNA *by L. R. McBride*
KONA LEGENDS *by Eliza D. Maguire*
LEAVES FROM A GRASS HOUSE *by Don Blanding*
PARADISE LOOT *by Don Blanding*
PETROGLYPHS OF HAWAII *by L. R. McBride*
PLANTS OF HAWAII *by Fortunato Teho*
PRACTICAL FOLK MEDICINE OF HAWAII
by L. R. McBride
STARS OVER HAWAII *by E. H. Bryan, Jr.*
THE STORY OF LAUHALA *by Edna W. Stall*
TROPICAL ORGANIC GARDENING - Hawaiian Style
by Richard Stevens

HAWAIIAN ANTIQUITY POSTCARDS
JOHN WEBBER PRINTS